NEXT-GENERATION MILITARY TECHNOLOGY

ROBOTICS
IN THE MILITARY

by George Anthony Kulz

BrightPoint Press
San Diego, CA

BrightPoint Press

an imprint of ReferencePoint Press, Inc.
Printed in the United States

For more information, contact:
BrightPoint Press
PO Box 27779
San Diego, CA 92198
www.BrightPointPress.com

LIBRARY OF CONGRESS CATALOGING-IN-PUBLICATION DATA

Name: Kulz, George Anthony, author.
Title: Robotics in the military / by George Anthony Kulz.
Description: San Diego, CA: ReferencePoint Press, 2027 | Series: Next-generation military technology | Includes bibliographical references and index. | Audience: Grades 7–9
Identifiers: ISBN 9781678214241 (hardcover) | ISBN 9781678214258 (eBook)
The complete Library of Congress record is available at www.loc.gov.

CONTENTS

AT A GLANCE

- Robotics is the science of making and using robots, machines that are programmed to carry out tasks. The military uses robots.

- One of the earliest uses of military robots was during World War II (1939–1945). The German military used small remote-controlled explosive vehicles.

- Military robots have advantages for reconnaissance missions. They can go places that people cannot safely travel.

- Military robots keep soldiers safer in battle. They scout ahead and look for danger. They also defend soldiers nearby.

- There are many ways military robots protect people from harm. They go on search and rescue missions. They also bring supplies to people in need.

- Robots are designed and programmed to handle specific jobs in the military. Soldiers practice and train with them. This helps soldiers work with robots effectively.

- Military robots provide medical help for soldiers. They carry injured soldiers off the battlefield to a safe place.

- Robots can help build structures such as huts and bridges in dangerous places. They are also good at repairing soldiers' equipment.

SPECIAL MEDICAL ASSISTANTS

On the battlefield, an army medic sees people who need help. Some are hurt and cannot walk. One person cannot breathe well. The medic asks a helper for an **intubation** kit. The helper is not a human being. It is a four-legged robot. It rushes away to fetch the item. When it comes back, the medic uses the kit to help the person breathe.

Four-legged robots may someday help human medics care for injured soldiers on the battlefield.

The medic asks another robot to look for more injured people. The robot races away. It goes to the next injured person. It uses **sensors** to learn about their condition. It returns to tell the medic this information. This helps the medic know what to do next.

This battlefield is not real. It is part of a practice exercise. The exercise shows how robots and people can help save soldiers together. Army medics are excited about this work. It will be useful in saving more lives. With robots, fewer medics can help more people.

ROBOTICS AND THE MILITARY

Robotics is the science of making and using robots. Robots are programmed machines. They do jobs with little to no help

Soldiers practice working alongside a wheeled robot to evacuate wounded teammates.

from people. They do many tasks faster and better than humans can.

Robots are widely used in the military today. They help find and remove bombs. Robots carry supplies. They help move

The robots being tested today will be used in the conflicts of tomorrow.

soldiers to safety. And they can be used to fight enemies.

There are many benefits to using robots. As long as they have power, these machines never get tired. Robots can work in dangerous environments. Finally, they can reduce or prevent harm to human beings. The battlefields of the future are likely to include humans and robots working together.

CHAPTER ONE

MILITARY MACHINES

In 2024, retired US Army general Mark Milley talked about robots. He saw them being used in the US military. He said more robots would be used in the future.

He said, "We are on the cusp of seeing a pilot-less Air Force, a sailor-less Navy, a crew-less tank. It won't be 100 percent—you're still going to have pilots and sailors and soldiers, but you're going to see an

General Mark Milley is one of many military leaders who have spoken about the future of military robots.

MILLEY

increasing use of robots and robotic forces in the years to come."[1]

DEVELOPING MILITARY ROBOTICS

One of the earliest uses of military robots was in World War II (1939–1945). The German military used a vehicle they called the Goliath. It was a small, tank-like machine. A soldier controlled it remotely. Cables ran from a controller to the vehicle. The Goliath was loaded with explosives. When it got close to a target, the operator would blow it up.

These early machines paved the way for the robots of today. They were used as weapons. Goliath's main purpose was to blow up enemy tanks. It kept soldiers

British and US troops captured and studied many German Goliath robots.

Many military robots are tough, rugged machines.

farther from danger. This reduced risks to a military's own troops.

Military robots have changed a lot since then. Some are still controlled by people from far away. But today, some robots do tasks autonomously. This means they act without people controlling them. They use computers and smart software known as **artificial intelligence (AI)**.

THE BENEFITS OF MILITARY ROBOTICS

Robots do many jobs in the military. They help watch for enemies. They perform **reconnaissance**, or recon, missions in enemy areas. They move alongside soldiers. They can fight if needed. Robots also carry supplies and give medical help. They build and fix structures, too.

Military robots are designed to keep people safer. Robots go into places that are dangerous. They do jobs that are risky.

They perform search and rescue missions. Robots draw fire away from soldiers, too. All these things reduce the danger to human troops.

Robots do not get tired the way humans do. They can work as long as they have power. They do not need breaks. Robots do not feel pain like people do. Robots can be made stronger than people. They carry heavier loads. Armor protects them against bombs and bullets.

Robots do some tasks better than people. Their sensors help them see the world around them. Special cameras see things human eyes cannot. So do radar and sonar sensors. Soldiers review data from sensors to help them make decisions. Some robots use AI to find enemy targets.

Two-legged robots need a lot of testing to make sure they stay balanced.

Rough seas are among the tough conditions robots face.

The software helps the robot identify what it is looking at. When these systems work correctly, robots help prevent harm to innocent people.

THE CHALLENGES OF MILITARY ROBOTICS

There are challenges with using robots in the military. Robots may be good at going to dangerous places. But they can still be damaged. Heat and cold can make robots break down. Weapons can damage them. Robots walking on legs may fall over. Robots can be difficult and costly to fix.

Robots that use AI may have to make choices about keeping people safe. They might make a bad choice. This choice may accidentally harm people. Robots may also be **hacked**. These robots may be storing secret information. Having this information stolen by an enemy may put soldiers' lives at risk.

Robotics Careers

As military robots get more advanced, people will be needed to build and fix them. These workers must know electronics and computer programming. They should know how sensors help robots see. They need good project management skills. They must also follow military standards for building and using robots.

The US military is testing drones that will make decisions on their own.

CHAPTER TWO

RECON, SCOUTING, AND PATROL MISSIONS

The military uses robots to help with recon, scouting, and patrols. Robots go where soldiers cannot. They see better and move faster than people. Robots send information back to soldiers. Sometimes they are able to handle problems on their own.

One mission of these robots is to find enemies. They can move ahead of soldiers. If they are seen, enemies may try to shoot

Flying drones are key tools for recon.

at them. Once that happens, soldiers will know where the enemies are.

SURVEILLANCE AND RECON DRONES

Robots need to know what is around them to go on recon missions. People use their eyes to see. Robots can see in more ways.

Robots can send back video footage of an area before soldiers move in.

Robots have cameras to take pictures. They use sonar and radar. Robots can sense heat. They can hear noises and feel vibrations.

Recon robots move in many ways. Some use wheels, tracks, or legs to move on land. Some use wings or propellers to fly. Some even flap their wings like birds. Other robots swim using **thrusters**, water jets, or fins. Some crawl along the ocean floor on tank-like tracks. All of these options give recon robots a lot of flexibility. They can locate and track enemies wherever they are.

Recon robots have several benefits. First, they keep soldiers safe by going into dangerous areas. Second, they find and send back information right away. Soldiers get instant information.

Third, robots can study the data they collect to help soldiers understand the battlefield. Finally, robots can be programmed to handle new jobs if needed.

SCOUT ROBOTS

Scout robots also gather information. They go ahead of human soldiers. They check whether an area is safe. In 2020, expert Donald Sando said:

> *Enemies generally shoot first 80 percent of the time. We don't like those odds . . . we want to avoid the close fight if we can. If we can't avoid it, we want to enter it under conditions that are favorable to us.*[2]

Robots can help with this, keeping friendly soldiers out of harm's way.

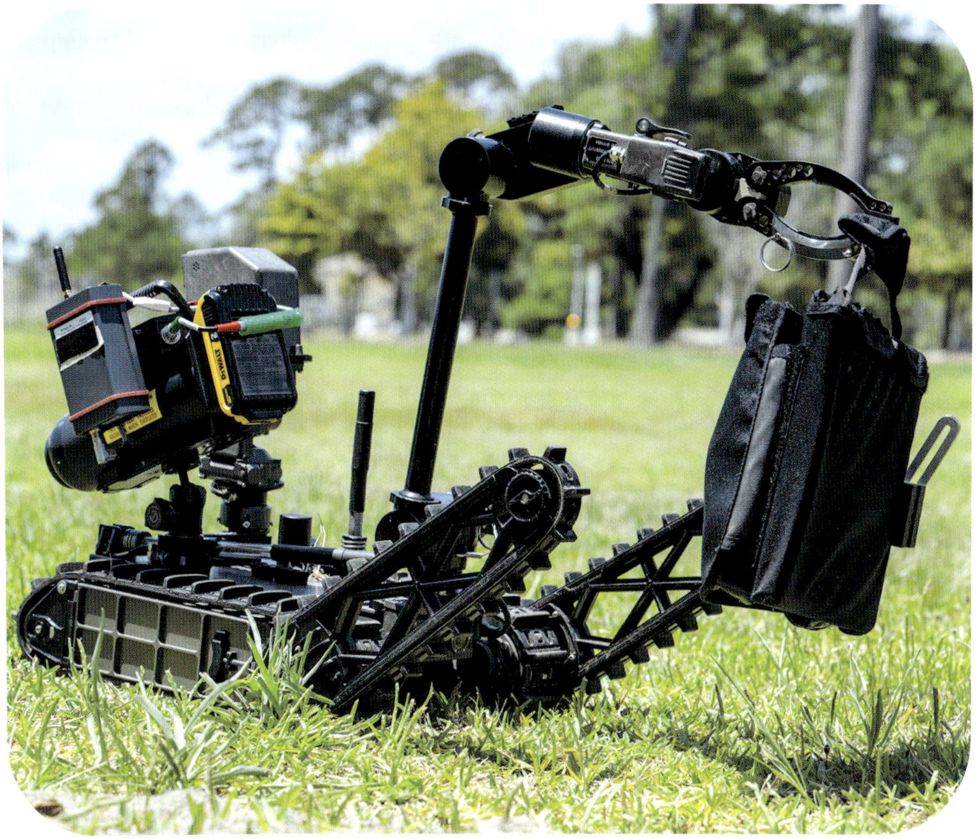

Robots can inspect suspicious bags that might contain bombs.

Scout robots are good at finding dangers that could hurt people. One danger is bombs. Robots find bombs and make them safe so no one gets hurt. They may also find enemy soldiers before people do. Robots can make the enemies shoot at them first instead of at people. Robots also

Each robot carries the sensors and other equipment needed for its missions.

help clear buildings so soldiers can go inside safely.

Scouting robots have many ways to tell soldiers what they find. They may send back a live video feed. They may send back data about temperature and air quality. Robots may send an alert that enemy troops are nearby.

ROBOTIC PATROL SYSTEMS

Robots are not only helpful to soldiers in the field. They also help protect bases. Patrol robots watch areas where soldiers stay. They keep soldiers safe. Robots do the patrolling rather than human guards. This reduces the danger to people.

Like other military robots, patrol robots use sensors. They look for strange activity. If they find danger, they tell nearby soldiers. They can watch and follow enemy soldiers. Some patrol robots have weapons. They may be allowed to shoot enemies they

US troops have trained for patrol missions alongside robots.

can see. Patrol robots can also work together in **swarms**.

Another benefit of patrol robots is that they save time. Human soldiers are freed up to do more important work. In 2023, Kimberly Rumph of the US Air Force spoke about this: "We save [time] which could be [used for] other activities requiring human logic and decision making."[3]

TALON

TALON is a remote-controlled robot. The US military first used it in 2001. Soldiers use it to find bombs and make them safe. It also detects gas, chemicals, and radiation. TALON is used for scouting missions, too.

SAVING LIVES AND FIGHTING BACK

One of the main jobs of military robots is to protect soldiers. They do this in many ways. Robots help prevent injuries by finding enemy bombs. They help treat wounded soldiers. Then they carry injured troops to safety.

Robots also help protect soldiers by attacking the enemy. They spot and destroy enemy equipment and vehicles. They may even fight against enemy troops on

Military drone pilots can both protect fellow soldiers and directly attack the enemy.

L3HARRIS

the ground. This helps keep human soldiers out of harm's way.

US Army leaders have spoken about robots working together with soldiers. In 2023, Secretary of the Army Christine Wormuth said,

Sending armored robots into battle can keep human troops safer.

These integrated formations will bring robotic systems into units alongside humans, with the goal of always having robots, not soldiers, make first contact with the enemy.[4]

SEARCH AND RESCUE

Robots help the military in search and rescue operations. They look for soldiers who are in trouble. They move the soldiers to a safe place.

Robots can search larger areas than soldiers can. They then tell soldiers where to find the people who need help. Some robots are made just for rescue jobs. They help save people by sharing important information fast.

Some ground robots used in Ukraine are large enough to carry troops to safety.

Some rescue robots have already gone into action on real battlefields. Ukraine used the MAUL remote-controlled ground robot in its war with Russia. In 2025, one of these robots rescued a Ukrainian soldier in enemy territory. He climbed inside, and the robot carried him to safety.

MEDICAL HELP

Robots help soldiers by giving medical help. Some robots help with surgery. Taking action quickly can improve a wounded soldier's chances of survival. Robots can get close to where soldiers are on the battlefield. Doctors who are far away control the robots and do the surgery. This puts the doctors at lower risk.

Robots check on injured soldiers by taking their vital signs and looking for wounds. They use cameras and sensors to check a soldier's heart rate. They see how the soldier is breathing or find injuries. Nathan Fisher works for the robotics

Robots have carried bags of medical supplies in tests.

company TATRC. He says, "The broad problem we're trying to solve . . . is how to provide [medical workers] with the most accurate and timely information."[5]

Sometimes robots help people in other ways. Robotic suits can help soldiers with spine injuries stand up and walk. They also help healthy soldiers pick up heavy things. These suits make it easier for them to walk long distances. Robotic arms and legs may take the place of real ones lost in war. They can improve life for veterans.

TRAINING ROBOTS FOR DANGEROUS JOBS

Military robots must be carefully designed and programmed to keep people safe. Robots learn to tell the difference between enemies and friends. They need to know when a place is dangerous. They need to be able to detect things like bombs, fires, floods, and poisonous air. The better they get at these things, the better they can protect people.

People design and program robots to help them carry out these jobs. A robot's brain is a powerful computer. Humans write programs that run on that computer. These programs help robots understand data from their sensors. Programs allow robots to recognize what they see and hear. They tell the robots what actions to take.

People also train robots by playing games called simulations. A robot's computer may be capable of learning from experience. A simulation lets the

Many hours of design and programming are needed before a robot can go into action.

robots practice how to act. They learn from mistakes and improve their skills.

Another good way for robots to learn is to let them work with soldiers. Robots are good at finding and studying information. People are good at relying on common sense and making choices. Soldiers can teach robots new things. Robots use this knowledge to do better at their jobs. This practice is useful for human soldiers, too. They learn how best to use robots on the battlefield. They figure out how to get the most out of these high-tech tools.

Helping During Disasters

Military robots are useful for more than just battles. Militaries often help during natural disasters. Robots can help during these crises. Ground robots may find people trapped in damaged buildings. Flying robots drop supplies into areas where they are needed. They take pictures of flooded areas. Robots in the water may rescue people from shipwrecks.

FIGHTING ENEMIES

Military robots do more than protect the soldiers on their side. They also find and fight enemies. Robots use their sensors to spot enemy soldiers. Some robots carry weapons such as guns or grenades. Others are made to explode near a target.

Some robots are controlled by people, but others are autonomous. They may make mistakes about whom to attack. In 2025, lawyer Bonnie Docherty wrote about the use of robots in war. She explained, "Machines do not have the capacity to . . . understand the consequences of their actions."[6] Militaries want to prevent these mistakes. They are working to make autonomous robots as smart as possible.

Ground robots with guns on them have been used in combat in Ukraine.

Robots do not just go after enemy soldiers. They can also strike other important things. Robots attack enemy tanks and planes. They may destroy radar systems and other equipment. This can make it harder for the enemy to see friendly forces. Robots also destroy vehicles carrying supplies.

Robots can even fight back against enemy robots. They do this by blocking the robots' signals. This is called jamming. When signals are jammed, enemy robots may not be able to work right. For example, jamming may make flying robots crash. It may even make robots attack their own side.

HELPING SOLDIERS IN THE FIELD

Robots help soldiers in ways that do not directly involve protecting them. They transport equipment and weapons. They also repair and build things that soldiers need to do their jobs.

On August 21, 2025, retired US Marine Corps general David Berger spoke about the importance of robots for these jobs. He said, "Do I need a human to drive that truck or fly that aircraft . . . or can we actually

A robot called the Mission Master carries cargo into dangerous areas.

US Marines have tested the delivery of lightweight cargo using small flying drones.

do it better, faster, 24/7, with [robots]? The answer is, yes, we absolutely could."[7]

TRANSPORT ROBOTS

Moving supplies and equipment is very important for the military. Transport robots carry things that are too heavy or hard for soldiers to move. These robots travel long distances. They cover rough ground. They help soldiers get supplies to challenging locations.

Transport robots may carry weapons and extra bullets to soldiers. They also bring food and water. They carry extra armor and communication equipment. This can free up human soldiers to do other work.

Sometimes robots help make sure soldiers' equipment works. Transport robots carry spare batteries. They also bring fuel for vehicles.

ROBOT BUILDERS AND REPAIR ROBOTS

Military robots help build bases and other things to keep soldiers safe. They use 3D printing technology to build structures out of simple materials. They build huts that soldiers can live in. They build walls to block roads or protect people. They also build bridges so soldiers can cross over gaps. When robots do this work, soldiers do not have to do all the hard building. They just need to know how to tell the robots what to do.

Robots can also shape metal. This helps people make replacement parts for military equipment and vehicles. Robots can quickly make a new part. Soldiers get their vehicles and equipment fixed and ready to use much sooner.

Robots do more than just build and fix things. They also help soldiers talk to each other. When robots are placed in different spots, they can pass messages

Quickly building bridges is a crucial job for ground troops. Robots could someday help with this important task.

AIR, SEA, AND GROUND ROBOTS

XQ-58A Valkyrie

Autonomous Low-Profile Vessel

Strategic Quad for Reconnaissance & Logistics

The military is testing and using robots in the air, at sea, and on the ground.

to each other. Each robot acts as a **relay**. It gets a message and passes it along to the next robot in a chain. Getting good information quickly is important for military missions.

SUPPORTING HUMAN TROOPS

Robots in the military provide many ways to keep soldiers safe. They go into dangerous places. They help save soldiers' lives. And they directly attack the enemy.

However, it is important for military robots to be used carefully. Human lives will be at risk. People should always be involved. This will help make sure robots do not make deadly mistakes.

In the future, humans and robots will work side by side. General James Rainey

Training will be key to using robots effectively alongside human troops.

of the US Army says the goal is for robots to do support jobs. These are the kinds of tasks they do best. General Rainey says, "The most precious weapons system we have still is the U.S. Army soldier. How do we offload risk from our humans onto robots?"[8]

SMET: The Robotic Mule

SMET is a robot with six wheels. It carries heavy things for the US Army. It is remote-controlled. It holds up to 1,200 pounds (540 kg). The robot can travel up to 62 miles (100 km). It can go 27 miles per hour (43 kmh) on paved roads or about half that speed on unpaved roads.

GLOSSARY

artificial intelligence (AI)

the ability of a computer to do things that usually require human intelligence to do

hacked

gained access to a computer without permission in order to harm it or change its function

intubation

the process of putting a tube through someone's nose or mouth down into their windpipe to help them breathe

reconnaissance

to gather information about a person, place, or thing before taking action

sensors

devices that pick up changes in the environment, such as light, heat, sound, or motion, and change them into data that a machine can understand

swarms

large groups of objects, animals, or people moving together

thrusters

devices that push air or water to help vehicles move

SOURCE NOTES

CHAPTER ONE: MILITARY MACHINES

1. Quoted in Cate Burgan, "Gen. Milley: U.S. Military on 'Cusp' of AI-Powered Operations," *MeriTalk*, June 17, 2024. www.meritalk.com.

CHAPTER TWO: RECON, SCOUTING, AND PATROL MISSIONS

2. Quoted in Sydney J. Freedberg Jr., "Army Wants Smaller Brigades, Stronger Divisions & Lots of Robots," *Breaking Defense*, November 6, 2020. https://breakingdefense.com.

3. Quoted in Jared Keller, "Robots Dogs Are Taking Over the US Military," *Tech & Tactics*, March 1, 2023. https://taskandpurpose.com.

CHAPTER THREE: SAVING LIVES AND FIGHTING BACK

4. Quoted in Stew Magnuson, "AUSA News: Army Envisions Robots Taking 'First Contact' in Conflicts," *National Defense*, October 9, 2023. www.nationaldefensemagazine.org.

5. Quoted in Paul Lagasse, "TATRC's VISTA Casualty Detection and Assessment Systems to Be Tested at Project Convergence Capstone 5," *US Army Medical Research and Development Command*, February 26, 2025. https://mrdc.health.mil.

6. Quoted in "'Killer Robots' Threaten Human Rights During War, Peace," *International Human Rights Clinic*, May 12, 2025. https://humanrightsclinic.law.harvard.edu.

CHAPTER FOUR: HELPING SOLDIERS IN THE FIELD

7. Quoted in Allyson Park, "Just In: U.S. 'Way Behind' Using Autonomy, Robotics for Logistics," *National Defense*, August 21, 2025. www.nationaldefensemagazine.org.

8. Quoted in Jen Judson, "US Army Developing Integrated Formations of Robots and Humans," *Defense News*, October 9, 2023. www.defensenews.com.

FOR FURTHER RESEARCH

BOOKS

Lisa Idzikowski, *Military Robots*. Lerner Publications, 2024.

George Anthony Kulz, *AI in the Military*. BrightPoint Press, 2027.

Lela Nargi, *Curious About Military Robots*. Amicus Learning, 2024.

INTERNET SOURCES

"Robot," *Britannica Kids*, n.d. https://kids.britannica.com.

"Ukrainian Startups Are Creating a Low-Cost Robot Army to Fight Russia," *NPR*, July 15, 2024. www.npr.org.

"Will Robots Someday Fight Our Wars?" *PBS*, n.d. https://pbs.org/newshour.

WEBSITES

DoD STEM

https://dodstem.us

DoD STEM provides resources on military-related STEM topics, including robotics, cybersecurity, and artificial intelligence.

FIRST Free Education Resources

www.firstinspires.org/resources/library/free-education-resources

The website for FIRST provides free resources related to robotics, such as courses, activities, and news.

RoboMasterminds

www.robomasterminds.org

RoboMasterminds is funded by the National Defense Education Program and provides information about STEM activities, workshops, and internships.

INDEX

IMAGE CREDITS

Cover: Sgt. Adel Pacheco Alvarez/DVIDS
5: Kenny Jones/US Navy/DVIDS
7: Cameron E. Parks/US Army/DVIDS
9: Sgt. Charlie Duke/US Army/DVIDS
10: Alex Hillman/DVIDS
13: Spc. Dustin D. Biven/US Army/DVIDS
15: © US Marine Corps
16: Airman 1st Class Samantha Melecio/US Air Force/DVIDS
19: John F. Williams/US Navy/DVIDS
20: Mass Communication Specialist 2nd Class Deven Fernandez/US Navy/DVIDS
23: Spc. Alyssa Robertson/US Army/DVIDS
25: Spc. Aliyah Vivier/US Army National Guard/DVIDS
26: LCpl Logan Beeney/US Marine Corps/DVIDS
29: Senior Airman Hussein Enaya/US Air Force/DVIDS
30: Sgt. 1st Class John Carkeet/US Army/DVIDS
32: Pfc. Rhita Daniel/US Marine Corps/DVIDS
35: Connor Taggart/US Marine Corps/DVIDS
36: Thomas Sakell/DVIDS
38: © Drop of Light/Shutterstock Images
40: Cameron E. Parks/US Army/DVIDS
43: Sgt. 1st Class Eric Maldonado/Puerto Rico National Guard/DVIDS
46: © Anton_AV/Shutterstock Images
49: Lance Cpl. Tucker Mocan/US Marine Corps/DVIDS
50: Kevin Ray J. Salvador/US Marine Corps/DVIDS
53: Pfc. Tomas J. Arce/US Army/DVIDS
54 (top): Master Sgt. John McRell/US Air Force/DVIDS
54 (middle): Sgt. Mary Torres/US Marine Corps/DVIDS
54 (bottom): Petty Officers 3rd Class Carmen Caver/US Coast Guard/DVIDS
56: Master Sgt. Amanda C. Geiger/US Air National Guard/DVIDS

ABOUT THE AUTHOR

George Anthony Kulz is a member of the Society of Children's Book Writers and Illustrators and has taken courses at the Institute of Children's Literature and the Gotham Writers' Workshop. He writes for young people and adults.